THIS BOOK BELONGS TO

———————

即将启程上海探索之旅

多彩城市
涂出你的上海
City of Colors
Bring Shanghai's Architecture to Life!

阿奇上海 著
ArchiShanghai

同济大学出版社
Tongji University Press

To Shanghai — our beloved city
致上海——我们心中永远的港湾

城市行走书系
策划：江岱、姜庆共

多彩城市：涂出你的上海
文字：阿奇上海
绘图：阿奇上海
地图：孙晓悦
封面插图：阿奇上海

责任编辑：罗璇
书籍设计：孙晓悦

鸣谢：
张丽、阮仪三、华霞虹、
斯文·赛拉诺、杜超瑜
哈佛大学中美国际交流协作机构
中国大智汇创新研究挑战赛

CityWalk Series
Curator: Jiang Dai, Jiang Qinggong

City of Colors: Bring Shanghai's Architecture to Life!
Text: ArchiShanghai
Illustration: ArchiShanghai
Map: Sun Xiaoyue
Cover Illustration: ArchiShanghai

Editor: Luo Xuan
Book Designer: Sun Xiaoyue

Acknowledgements:
Zhang Li, Ruan Yisan, Hua Xiahong,
Sven Serrano, Du Chaoyu
Havard Association for US-China Relations (HAUSCR)
China Thinks Big

目录

序 ················· 8
涂色之前，请先阅读 ················· 10

多彩城市

　　1933 老场坊 ················· 12
　　百乐门舞厅 ················· 14
　　大光明电影院 ················· 16
　　丁香花园 ················· 18
　　国际饭店 ················· 20
　　鸿德堂 ················· 22
　　华东政法大学 ················· 24
　　江湾体育场 ················· 26
　　绿房子 ················· 28
　　马勒别墅 ················· 30
　　上海展览中心 ················· 32
　　田子坊 ················· 34
　　武康大楼 ················· 36
　　小红楼 ················· 38
　　新天地 ················· 40
　　徐家汇天主堂 ················· 42
　　杨树浦水厂 ················· 44
　　豫园 ················· 46
　　朱家角 ················· 48

趣味问答 ················· 50
推荐阅读 ················· 54
一本涂色书的诞生 ················· 55
致谢 ················· 58

Table of Contents

Preface ················· 9
Before You Start... ················· 11

City of Colors

　　1933 Old Milfun ················· 12
　　Paramount Hall ················· 14
　　Grand Theatre ················· 16
　　Lilac Garden ················· 18
　　Park Hotel ················· 20
　　Fitch Memorial Church ················· 22
　　East China University of
　　　　Political Science and Law ················· 24
　　Jiangwan Sports Stadium ················· 26
　　The Green House ················· 28
　　Moller Villa ················· 30
　　Shanghai Exhibition Center ················· 32
　　Tianzifang ················· 34
　　Normandie Apartments ················· 36
　　Little Red House ················· 38
　　Xintiandi ················· 40
　　St. Ignatius' Cathedral ················· 42
　　Yangshupu Water Plant ················· 44
　　Yuyuan Garden ················· 46
　　Zhujiajiao ················· 48

Time for an ArchiQuiz! ················· 51
Recommended Readings ················· 54
The Making of *City of Colors* ················· 56
Many Thanks To... ················· 59

序

阿拉桑海宁

这本关于城市历史的有趣绘本，起源是一群在上海生活的国际高中生的身份困惑。因为对于"你是哪里人"这样一个问题，同学们不假思索的回答中鲜有"中国人"，更不要说"上海人"。

这本无可厚非。

因为他们拿着外国护照，接受着西式教育，操着流利纯正的美式英语，毕业后会去国外追梦，上海不过是少年时的驿站。

然而，有八位年轻人，希望回答得不一样。

借"中国大智汇"比赛的机会，他们想让自己，也想让更多中国和外国的同龄人了解上海。受到一次舞会活动场地在中西合璧的萨莎花园的启发，他们想用"历史建筑"这个切入口，做一本建筑涂色本，像《秘密花园》一样吸引人。

他们建起"阿拉桑海宁"微信公众号，他们搜索历史，行走城市，拍摄照片，绘制图画，采访专家，采访同学，分发问卷，用自己的疑问和反思一点点充实着小站……终于，在指导老师的朋友圈看到了他们获大赛亚军的欢乐合影。

但他们对这座城市的热情已经发酵。上海书展上，在同济大学出版社展台，"阿拉桑海宁"让城市冰凉厚重的砖石在老少读者的笔下变得缤纷而温暖。

又是半年，马勒别墅化成了梦境中的移动城堡，透过装饰艺术风格标志构成的绚丽银幕，看到堂皇的百乐门舞厅，巨大的自来水管和龙头托起杨树浦水厂，上海展览中心的尖塔和五角星融入童话的森林……十九处不同年代，风格迥异的历史遗产，化作一幅幅充满想象力的画页，仿佛一张张来自城市的邀请函。

跟自然花草相比，城市建筑的形态和色彩要乏味很多。如果不能突破三维的限制，很容易变成照片的翻版，即使涂得五颜六色，也无法带给读者视觉的新鲜感和创作的成就感。除了从建筑外观和历史背景方面打开脑洞外，更富有挑战性的是将建筑元素变形重组后形成装饰性平面，比如那幅如棱镜折射，又似埃舍尔迷宫的画面，不仅巧妙融合了建筑立面装饰和室内坡道，还传神地展现了1933老场坊独有的空间体验，层层飞桥间，人头攒动……

是啊！在城市的历史中加入个人成长的故事，温度自然不会退却。

八位年轻人这一年的努力，足以让他们自己，也让喜欢用自己的方式阅读这座城市的人们，情不自禁说出一句，"阿拉桑海宁！"

华霞虹

同济大学建筑与城市规划学院 副教授
2017年4月25日

Preface

I'm Shanghainese

This fascinating coloring book on Shanghai's history originates from a group of international high school students' identity crises. Although they have studied and lived in Shanghai for several years, when asked "where are you from," many of their immediate answers rarely included "China," let alone "Shanghai."

This should not come as a surprise.

Holding foreign passports, receiving western education, and speaking fluent American English, these students will pursue their dreams abroad after graduating. Shanghai is just a post in their life journeys.

However, eight teenagers wished for a different answer. Seizing the opportunity provided by China Thinks Big, they wanted to let not only themselves, but also peers from both China and abroad understand Shanghai better. Inspired by the story of their winter ball venue, Sasha's Garden, which integrates Chinese and Western architectural styles, they decided to showcase historically significant architecture in the form of a coloring book that is as enchanting as *Secret Garden*.

They created the WeChat subscription account "ArchiShanghai" (the Chinese name Alasanghaining means "I'm Shanghainese" in Shanghainese dialect); they researched the history behind Shanghai architecture, toured the city, took photos, produced illustrations, interviewed experts and peers, distributed surveys, and expanded their account with questions and reflections... Finally, I saw a lovely photo of them winning second place at the competition through their supervisor's WeChat moments.

Even though the competition ended, their passion towards Shanghai has already taken root, and they have continued to make more out of ArchiShanghai. For example, at Tongji University Press's booth at the Shanghai Book Fair, ArchiShanghai helped readers transform the heavy cold bricks and stones of Shanghai architecture into bursts of color and warmth.

And after half a year of imagination, Moller Villa turned into a fantastical castle found only in dreams, Paramount Hall's extravagant dance floor transferred onto a vibrant screen containing art-deco elements, lengthy hoses and a gigantic water tap held up the Yangshupu Water Plant, and the spire and star of Shanghai Exhibition Center merged into a forest full of fairytales. Nineteen historical legacies of different times and diverse styles have morphed into pages of delightful imagination, inviting readers to explore Shanghai.

Compared with plants in nature, architecture is dull in forms and colors. Without breaking the conventions of three-dimensional images, no matter how many colors are applied, the illustrations are still replicates of ordinary photos. To guarantee the readers with visual novelty and a sense of accomplishment, the designer should not only draw inspiration from the buildings' appearance and historical background, but also reconstruct architectural elements into graphic artwork, which is even more challenging. For instance, the piece that resembles reflective prisms and Escher's maze cleverly infuses exterior ornaments with interior ramps, vividly reproducing the spatial experiences unique to 1933 Old Millfun — between stories of bridges, clusters of silhouettes appear in and disappear out of sight...

Of course. With a touch of personal growth added into a city's history, passion will always remain.

With one year of effort, these eight teenagers have certainly allowed themselves, as well as those who enjoy interpreting this city in their own ways, to feel confident enough to shout out, "I'm Shanghainese!"

Huaxiahong

Associate Professor, CAUP, Tongji University
April 25, 2017

涂色之前，请先阅读

在当今的上海，矗立于陆家嘴的摩天大厦群已成为其高速发展的标志，衡山路一带灯红酒绿、华光流彩的繁华夜生活也成了其国际化的象征；而另一方面，那些带着尘封记忆的老式弄堂又向我们诉说着这些街道楼宇所目睹的历史变迁。中西合璧，新旧结合——上海的建筑文化是如此独树一帜，其背后所承载的历史也是如此厚重：它既背负了西方资本帝国主义入侵的辛酸，又孕育着中国现代高速发展崛起的希望。然而，太多从其他城市和国家来的人，甚至是上海本地的年轻人，都对这些文化遗产疏于了解。我们想凭借自己的热情和决心尝试改善这一问题，《多彩城市》由此而生。

作为国际学校的高三学生，阿奇上海的成员都将在一年内离开上海，去国外追寻各自的梦想，所以相信这本书会是我们报答上海——这座已经成为我们家乡的城市——最好的方法。其实，在制作《多彩城市》的过程中，我们自己也学到了许多关于上海建筑的知识，并再次着迷于她所承载的丰厚历史文化。我们由衷地希望，拿着这本书的您也能够如此。

在本书中，您会找到以下内容。

——原创建筑手绘：各位可以用还原建筑原貌的方式涂色，也可以尽情挥洒创意天赋，涂出想象中的上海建筑；

——小段文字：解说本书创作者绘制每座建筑时的灵感来源；

——趣味小知识：介绍每座建筑独一无二的历史与特征；

——趣味小图标：放大建筑细节，反映场所特点，或提示历史背景，也可用来涂色；

——当然还有必不可少的涂鸦乐趣！

话不多说，赶快拿起画笔，去领略上海建筑的无穷魅力吧！

阿奇上海
2017 年 2 月

Before You Start...

Metropolitan skyscrapers and dazzling skylines are pictured in most people's minds when they think of Shanghai's architecture. In reality, Shanghai is home to a diverse body of architecture — a result of Shanghai's history as a prosperous international trade hub that welcomed foreigners to incorporate architectural elements from abroad into the construction of local buildings. Today, however, more and more foreigners, migrants, and even local Shanghainese fail to appreciate these cultural relics as they experience the fast pace of modern city life. Yet, we believe such an important part of Shanghai's cultural identity deserves to be understood and passed on.

As international high school students who will be leaving Shanghai in less than a year to pursue our dreams abroad, we consider this coloring book a gift that we want to give to the city that has become our home for so long. Through its creation, we were able to deepen our knowledge and understanding of Shanghai, and we sincerely hope this book can do the same for you.

In this coloring book you will find...

- Creative interpretations of Shanghai's architecture for you to bring to life with color. (You can either color them as realistically as possible or color them however you like using your imagination!)
- Short paragraphs explaining what inspired our artistic designs.
- Fun facts for you to learn.
- Icons with captions that portray further details or characteristics of the architecture. (You can also color these in!)
- And of course, coloring delight!

Without further ado, you will now embark on a journey to explore the magical city of Shanghai and learn about its unique architecture!

Have Fun!

ArchiShanghai
February 2017

1933 老场坊

来到1933老场坊，高低错落、蜿蜒盘旋的廊桥和楼梯使人仿佛置身于迷宫之中，外墙密密麻麻的圆形镂空装饰更是令人眼花缭乱。有谁能想到，这座极富艺术气息的建筑曾是上海工部局的宰牲厂！可要小心，在忙着涂色的同时，千万不要迷路哦！

1933 Old Milfun

If you decide to visit the 1933 Old Milfun, you should better get yourself a map of its interior because this building has a unique, maze-like structure that mesmerizes its visitors. Various people such as artists, tourists and businessmen may come here to enjoy a variety of activities like conferences, art exhibitions, fashion shows, etc. This illustration gives you an impression of its complex structure, with an emphasis on the gallery bridges and spiral staircases that connect different parts of the building.

虹口区溧阳路611号（近沙泾路）
No. 611 Liyang Rd. (near Shajing Rd.), Hongkou District

- 1933年竣工

- 上海公共租界工部局出资，英国设计师设计

- 外方内圆，伞形柱，无梁楼盖，牛道、旋梯宛若迷宫

- 外墙满是圆形的镂空装饰；原上海工部局宰牲厂，现为举办艺术展览的创意园区

- Completed in 1933

- Commissioned by Shanghai Municipal Council; designed by British architects

- Concrete structure; spiral staircases

- This building was originally constructed as a slaughterhouse of the Shanghai Municipal Council. By 1945, it became the largest slaughterhouse in the Far East and provided two-thirds of fresh meat in Shanghai. This building is now a "creative industry zone", hosting restaurants, cafes, exhibitions, and other creative activities.

百乐门舞厅

闪烁陆离的霓虹灯光、慵懒舒缓的爵士乐曲、玲珑曼妙的摇曳身姿——百乐门舞厅一直以来都是老上海灯红酒绿、纸醉金迷的夜生活的象征。如今的百乐门在重新修缮后，还添加了酒吧、餐馆、影院等多种综合性娱乐设施。虽然融入了不少现代特色，它却仍保留了一股子怀旧气息，仿佛在向人们诉说着七十年前的繁华与辉煌。从大屏幕里望去，可见那光洁的舞池地板、整齐排放的餐桌和幕布遮掩下的舞台。一曲终了，下一场精彩的表演即将开始。你，准备好了吗？

Paramount Hall

Want to go to a fancy place for a movie? Visit the extravagant theatre at Paramount Hall! The Hall is a symbol of Shanghai nightlife; it includes various entertainment facilities like the bar, restaurant, and dance floor, which are featured in the movie screen in this illustration. Are you ready to indulge yourself in some old-fashioned fun?

- 1932 年竣工

- 中国商人顾联承等营建，建筑师杨锡镠设计

- 立面强调垂直线条，转角主入口处上方有竖向长条窗，顶部高耸圆柱形霓虹灯塔。舞池地板用汽车钢板支托，跳舞时会产生晃动的感觉

- "远东第一乐府"，上海爵士乐的起源地，首席乐队"金杰米"曾在此演奏《夜来香》等著名作品；"百乐门"是英语"Paramount"的音译，有"卓越"和"千变万化"之意；为了便于跳舞，舞厅无一根立柱；因灯光和玻璃地板，百乐门舞厅曾被称为"玻璃世界"，可容数百人共舞的舞厅又被称为"千人舞厅"；新中国成立后，主体及附属建筑改为红都戏院和商场，后又改为红都电影院，演出越剧、沪剧，放映电影

- Completed in 1932

- Commissioned by Chinese merchant Gu Liancheng, et al.; designed by Chinese architect S.J. Young (Yang Xiliu)

- Dance floor supported by steel plate of cars to produce rocking movements; large glass light tower on the top floor

- The Paramount Hall is the origin of Shanghai's jazz music led by the first Chinese jazz band Jimmy King. It also provided creative services like number card that notify chauffeurs when the car owners were leaving. In 1936, Charlie Chaplin and his wife danced here!

静安区愚园路 218 号（近华山路）
No. 218 Yuyuan Rd. (near Huashan Rd.), Jing'an District

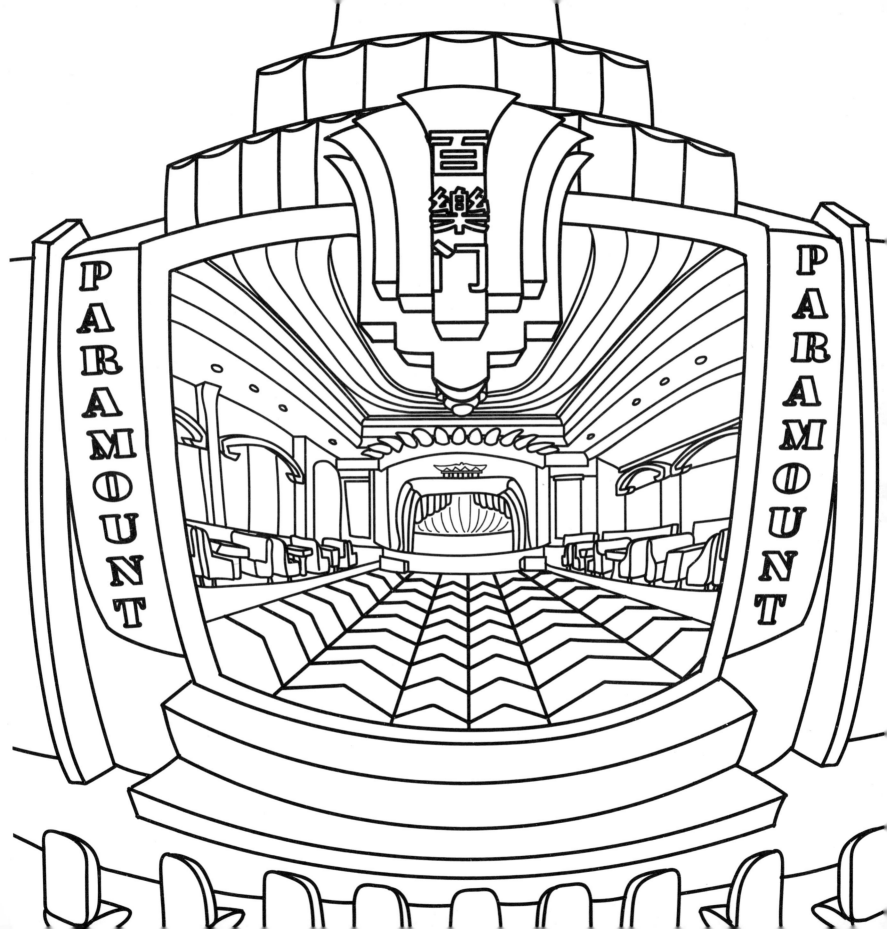

大光明电影院

在繁忙的现代都市中，电影已成为我们必不可少的生活调味剂。八十多年里，这座电影院见证了中国影视行业的发展历程。奶黄色的墙体上，横竖交错的线条和大片的玻璃产生强烈的对比，高耸的玻璃灯柱上"大光明影院"几个英文字母参差错落，为建筑整体带来独特的艺术感。图中左上方的胶卷则印着影院早期放映的两部电影《一夜风流》与《亮眼睛》中的人物，影院入口上方也张贴着《亮眼睛》的主演詹姆斯·杜恩的宣传海报。来吧，发挥你的想象力，为电影中那一个个天马行空、光怪陆离的世界添上你心目中的色彩！

Grand Theatre

The film reel fluttering in the sky showcases the main characters of the movies *It Happened One Night* and *Bright Eyes*, which were among the first movies screened in the Grand Theatre. The movie poster on the theatre also displays an actor who starred in *Bright Eyes*, James Dunn. The Grand Theatre itself features a white rectangular tower with the words "Grand Theatre" written on it that lights up nicely at night. Its exterior makes use of horizontal and vertical lines, which are emphasized through the comet-like pathway drawn in this illustration.

黄浦区南京西路 216 号（近黄河路）
No. 216 West Nanjing Rd. (near Huanghe Rd.), Huangpu District

- 1933 年重建

- 匈牙利建筑师邬达克设计

- 外观立面上横竖线条与体块交错，入口处有高达 30.5 米的方形半透明玻璃灯柱；室内采用弧线形的墙面、顶棚、灯光带等动感设计，具有现代感

- "远东第一影院"；梅兰芳为影院开张剪彩；第一家使用"译意风"（类似同声翻译）的影院

- Rebuilt in 1933

- Designed by László Hudec

- Intersection of horizontal and vertical lines on the exterior; 30.5-meter semi-transparent glass lamp post

- The Grand Theatre is often called the "first theatre of the Far East." It is also the first theatre in China to adopt simultaneous translation earphones.

丁香花园

丁香馥郁，香樟繁茂，巨龙盘踞，曲径通幽——坐落于华山路的丁香花园在传统园林中融入西式花园洋房的格调，处处弥漫着浪漫的情怀。图中出现在西式别墅上方的正是丁香花园传说中的主人——晚清北洋大臣李鸿章，背景则是园内特有的"一条蛟龙卧半园"的龙墙。

Lilac Garden

The Lilac Garden is a classic exemplar of Shanghai international culture, as it embodies elements of both traditional Jiangnan-styled gardens and Western garden houses. Here, one of the garden houses in Lilac Garden is personified by the famous Qing Dynasty minister Li Hongzhang, who was said to be the original owner of the Garden. The curved, scaled wall behind Li gives you a glance of the lengthy dragon wall in the Garden.

徐汇区华山路849号（近复兴西路）
No. 849 Huashan Rd. (near West Fuxing Rd.), Xuhui District

- 建造时间可能为1862年或1900年
- 传说由美国建筑师艾赛亚·罗杰斯设计
- 园区西北部为西式住宅，中部为大草坪，东南部为中式花园。花园中有湖，湖畔蜿蜒一道龙墙；湖上架九曲桥，通往湖心八角亭；亭上立金凤凰，与龙墙端头的龙首隔湖相望，呈龙凤戏水之意
- 住宅及花园20世纪为李鸿章之子李经迈居所，传说系李鸿章为七姨太丁香所建。如今经营特色上海菜和粤菜的申粤轩酒楼坐落于此

- Built in 1862 or 1900
- Said to be designed by American architect Isaiah Rogers
- Western garden houses; zigzag dragon wall; phoenix statue; zigzag bridge on the lake; lush flowers and trees
- The Lilac Garden was said to be named after Li Hongzhang's concubine called "Lilac" (Ding Xiang), to whom this garden was dedicated.

趣味问答 ArchiQuiz 3.C

国际饭店

来上海旅游的话，下榻国际饭店一定会是一个不错的选择！这座位于南京西路的酒店有着超过八十年的历史，也曾因其高度收获了"仰观落帽"的美誉。有心的你或许已经发现，右侧画面中央的方形图案和酒店大堂的天花板有着异曲同工之妙，周围环绕的四颗星也对应了酒店四星级的高质量服务水平。

Park Hotel

If you are visiting Shanghai, staying at Park Hotel will be a great choice! The Hotel had been the tallest building in Shanghai until the 1980s and remains a key architectural landmark here. In this illustration, the four repeating images of the Hotel exterior connect to form the interior ceiling structure. Hint: the four stars on the ceiling suggest that it is a four-star hotel!

黄浦区南京西路 170 号（近黄河路）
No. 170 West Nanjing Rd. (near Huanghe Rd.), Huangpu District

- 1934 年竣工

- 邬达克在纽约、芝加哥等地观摩美国摩天楼获得了设计灵感，并针对上海软土地基进行特殊结构技术处理，缓解了高层建筑的沉降问题

- 立面采取垂直线条划分；15 层以上呈阶梯状的塔楼；以深褐色面砖拼砌成富有韵律感的花纹，底部为黑色磨光花岗岩的饰面

- 四星级；"远东第一高楼"，保持上海建筑高度神话长达半个世纪；美国装饰艺术派风格高层摩天楼的翻版；中美长途电话通话典礼曾在此举行，宋美龄出席；楼顶旗杆的中心位置被定义为上海城市测绘的零坐标；著名华裔建筑师贝聿铭被国际饭店的设计打动而选择建筑专业

- Completed in 1934

- Designed by László Hudec

- Façade emphasizing on vertical stripes

- This four-star hotel was dubbed as "the first skyscraper of the Far East" in the 1930s. The opening ceremony of China-US long distance call service was held here. For the urban topographic survey in 1950, the center of Park Hotel was set as the origin of coordinate of Shanghai!

鸿德堂

你可曾见过中式的基督教堂？位于虹口区多伦路的鸿德堂就是这么一朵建筑奇葩。它打破了教堂惯有的西式建筑传统，转而采用中国古典建筑特有的斗拱和飞檐的样式，不禁令人称奇。我们将鸿德堂的外部大胆地置于其内部，不知有没有体现出那种特立独行的创新精神呢？

Fitch Memorial Church

Have you ever seen a Chinese-style Christian church? Unlike most other churches from this period, the Fitch Memorial Church (Hong De Tang) was built in the traditional Chinese temple style, subverting the custom of constructing churches in Western style. This illustration combines the interior — its ceiling and walls — with the exterior look of the Church. The clouds around the building and the big cross on top of it add a heavenly sense to the Church.

- 1928 年竣工

- 前身为美国北长老会在沪的第一所教堂"思娄堂"，1882 年由著名传教士费启鸿牧师设立；1925 年迁至虹口，建成后为纪念费启鸿牧师更名为"鸿德堂"

- 重檐式大屋顶，顶层为钟塔；屋顶与墙体间以斗拱相接；外墙以青砖砌成；仿木构架的暗红色混凝土圆柱，仿造中国传统寺庙立柱的色彩形式却以法国古典主义建筑常用的双柱形式出现；内部采用西式教堂高耸的空间

- 上海现存唯一的有中国殿宇特色的基督教堂；建造时任鸿德堂主持牧师的中国牧师陈金镛提出，教堂地处华人聚居区，应选用中国传统建筑风格；教堂以中国古典建筑样式去适应宏大的西式宗教空间要求，是中西融合的典型

- Completed in 1928

- Originally founded by American Presbyterian Mission (North); renamed "Hong De Tang", after moving to its present location, to commemorate Reverend George Field Fitch, whose Chinese name was "Fei Qi Hong"

- Multi-layered eaves on roof, covering a bell tower; "dougong" (wooden interlocked buttresses) connecting roof and walls; grey bricks and dark-red concrete columns on the exterior

- Fitch Memorial Church is the only existing Christian church in Shanghai built in traditional Chinese style. The unique style of this church was influenced by the Chinese priest Chen Jinyong at the time of construction.

虹口区多伦路 59 号（近四川北路）
No. 59 Duolun Rd. (near North Sichuan Rd.), Hongkou District

华东政法大学

笔直的线条，提醒着每一位法学人需公正无私；高悬的天平，象征着法律的神圣和不可亵渎——想必这也是华东政法大学的校徽所承载的意义，自创建以来勉励莘莘学子韬奋前行。（"韬奋"二字于华东政法大学也有着特殊的意义哦！）

East China University of Political Science and Law

Law, Justice, Equality. These are some ideas you may be reminded of when you see the huge scale towering over this building. Actually, the scale represents the spirit of East China University of Political Science and Law, a university well-known in politics and law throughout China. The building featured in this illustration is the university's Taofen Building. Notice the statue of Mr. Zou Taofen, after whom this building was named!

- 1879年美国籍主教施约瑟在苏州河南岸的梵皇渡码头创办圣约翰书院，依照美国大学形制办学，发展为圣约翰大学，即华东政法大学的前身

- 如今作为教学楼的韬奋楼，原名怀施堂，建于1894年；为两层砖木结构，结合中西风格；四面围合，外廊围以铸铁栏杆；外墙以青红砖相间砌筑；主入口顶部建有重檐屋顶的钟楼

- 邹韬奋（1895—1944）是中国近代著名出版家，1921年毕业于圣约翰大学。1926年起在上海担任《生活》周刊主编，曾将其发行量提升至逾15万份，创造了当时杂志发行的新纪录

- Founded by American bishop Joseph Schereschewsky, St. John's College, later known as St. John's University, was an institution that followed the American system.

- In 1952, East China University of Political Science and Law was founded here, one of the first higher institutions of politics and law in the PRC.

- Schereschewsky Hall (now Taofen Building) has wood-and-brick structure, red and grey brick walls, open-air galleries with cast iron railings, gable and hip roof with Chinese "butterfly" tiles, and a bell tower.

- After graduating from St. John's University, Zou Taofen (1895-1944) became Chief Edior of *Shenghuo Zhoukan*, a magazine based in Shanghai. He was one of the most famous publishers in modern China.

长宁区万航渡路1575号（近凯旋路）
No. 1575 Wanhangdu Rd. (near Kaixuan Rd.), Changning District

江湾体育场

　　如果你热爱运动，那么上海江湾体育场就是你绝佳的锻炼场所！体育场周围筑有高11米、可容纳数万人的二层看台，西侧由三座拱门构成的主入口也是其特色之一。奔跑吧，少年！在夕阳的余晖中，尽情地挥洒汗水，燃烧你的青春吧！

Jiangwan Sports Stadium

　　The soccer ball in the sky hints at the huge field behind the Stadium's main entrance as depicted in this illustration. As you come to the Stadium to play sports like badminton and tennis, you can appreciate the intricate patterns found on the arches of the architecture.

杨浦区国和路346号
No. 346 Guohe Rd., Yangpu District

- 1935年竣工

- 中华民国国民政府《大上海计划》的重要组成部分，1933年上海市政府为迎接民国第六届全运会下令建造，由中国建筑师董大酉设计

- 包括体育场、体育馆、游泳池三大建筑，外立面为清水红砖墙；体育场的椭圆形看台为长达千米的环形建筑，东、西司令台由人造汉白玉筑成，西司令台正门立面设3座高8米的拱形大门，建筑上部和檐部点缀传统式的小构件和线脚，其左右顶端设置古铜色大鼎各1座，供插放火炬

- "远东第一体育场"；看台有34个出入口，只需5分钟观众便可全部离场；被誉为现代化中国体育建筑；1935年曾举办民国第六届全国运动会

- Completed in 1935

- Commissioned by Shanghai Government for the Sixth National Games of the Republic of China; designed by Chinese architect Dayu Doon (Dong Dayou)

- Three main buildings— sports stadium, gymnasium, swimming pool

- The Jiangwan Sports Stadium is often called "the first sports stadium of the Far East." It has 34 entrances and can evacuate the audience in just five minutes. It also hosted the Chinese National Games in 1935.

绿房子

地处上海静安区市中心的吴同文住宅由著名匈牙利设计师邬达克设计。在建成时，它是全上海最豪华的洋房之一。因其外墙贴满了绿砖，它又常被老上海称为"绿房子"。院内林木茂盛，甚至在屋顶上也设有花园，养殖花鸟。咦，有两只淘气的小猫猫好像躲起来了，快帮吴同文先生把它们找回来！

The Green House

This building was the luxurious mansion of the Chinese tycoon D.V. Woo. It has the nickname "the Green House" due to its green-tiled exterior. The greenery in this illustration highlights the stylish garden on the roof of D.V. Woo's residence. Oh, two little cats are hiding here! Can you help Mr. Woo find them?

静安区铜仁路 333 号
No. 333 Tongren Rd., Jing'an District

- 1937 年竣工

- 邬达克设计

- 首层中间架空作汽车道，将建筑分为两部分——社交空间与主人用房；圆弧形阳光房通高四层，与层层递退的大露台形成纵横对比；弧形大楼梯连接花园与露台

- "远东最大最豪华的住宅之一"；上海首家装电梯的私宅；海派女作家程乃珊小说《蓝屋》的原型

- Completed in 1937

- Designed by architect László Hudec

- Comprised of two sections — one for guest reception and the other for residence; cylinder-shaped sunroom; terraces full of smooth curves; curved stairways

- D.V. Woo's Residence is often called "one of the largest and most luxurious residences in the Far East." It was the first private residence with an elevator in Shanghai. This building is also the prototype of the novel *The Blue House*, written by Cheng Naishan, a Shanghainese female writer known for her Shanghai School (haipai) writing style.

马勒别墅

　　马勒别墅宏伟而梦幻的外形总能让人想起童话书中的城堡。传说中，别墅的设计灵感来自英国船业大亨艾瑞克·马勒心爱的女儿的一个梦，这座建筑由此染上的奇幻色彩便是我们将马勒别墅打造成一座移动城堡的原因。悬浮的岛屿，缠绕的锁链，盘旋的巨龙——这一切都昭示着一段魔幻征途即将启程。试试能不能找出被囚禁的小公主？她正等待着你去拯救哦！

Moller Villa

Moller Villa's Scandinavian castle-style exterior reminds passersby of the dream-like mansions in fairytales. Legend has it that Eric Moller, a British merchant, built this storybook building for his daughter who dreamed of a castle like those in Andersen's fairytales. Here, the building fictionally rests on a floating island tied by steel chains to highlight its fantastical atmosphere. Can you spot the little princess looking out the window? She's waiting for you to rescue her!

静安区陕西南路 30 号（近延安中路）
No. 30 South Shaanxi Rd. (near Middle Yan'an Rd.), Jing'an District

- 1936 年竣工

- 英籍犹太富商马勒投资，华盖建筑事务所设计

- 墙面凹凸多变、棱角起翘，屋面高尖陡直，主体建筑顶部矗立着高低不一的两个四坡顶；花园围墙以黄绿色中式琉璃瓦压顶，门口放置石狮子；室内装修酷似豪华游轮；走廊上的木雕图案描绘船队在海上的情景（如船舵、海浪、海上日出、海上灯塔等）

- 传闻马勒因其女儿梦见拥有了一座"安徒生童话般的城堡"，便委托建筑师根据女儿梦中所见设计别墅；花园一角有青铜马，纪念马勒曾靠买马票中彩而发家

- Completed in 1936

- Commissioned by British shipping tycoon Eric Moller; designed by Allied Architects

- Pointy roofs, rugged walls, stained glass; wood carvings that depict shipboard scenes, like steer, anchor, waves, in the corridors

- The Moller Villa is often called a "Scandinavian fairytale style" building due to its castle-like exterior. Legend has it that this building was based on a castle that appeared in Eric Moller's daughter's dream.

上海展览中心

群鸟、仙鹤、驯鹿、棕熊——放眼望去，上海展览中心仿佛变成了童话中的动物王国、自然天堂！这样的设计正呼应了展览中心中那面刻满草木动物的外墙，充满了艺术气息。除此之外，上海展览中心还极富历史意义——那嵌于建筑最顶端的红五星，正是展览中心曾为中苏友好大厦最好的见证！

Shanghai Exhibition Center

You may wonder why an exhibition center is surrounded by what looks like an animal kingdom or an exquisite garden. Actually, the animals and trees mirror the relief sculptures on the exterior of the Shanghai Exhibition Center. Another key feature is the red star on the spire of the building, reflecting its history as the Sino-Soviet Friendship Building.

静安区延安中路 1000 号（近铜仁路）
No. 1000 Middle Yan'an Rd. (near Tongren Rd.), Jing'an District

- 1955 年竣工
- 新中国成立后不久，为介绍苏联经济文化成就、举办大型展览而建设相应展馆；方案在中苏建筑专家合作下产生，中国建筑师陈植、张乾源等参与设计
- 大厦坐北朝南，正南为广场，有音乐喷泉；主楼矗立正中，上竖镏金钢塔；大厦展厅及附属建筑，层层往后延伸；内有 40 多个大型展厅

- Completed in 1955
- Built to memorialize Sino-Soviet relations; designed in cooperation by Soviet architects and East China Construction Engineering Bureau
- Square with music fountain; central spire topped by a red star; relief sculptures on various parts of the building
- Used to be one of Shanghai's commanding grounds due to its tall spire; renamed "Shanghai Exhibition Center" in 1984

田子坊

在泰康路街头，可爱的一熊一虎笑眯眯地欢迎你来到以"小资"情调著称的田子坊！作为沪上最能体现海派特色的场所之一，田子坊既保留了传统里弄的生活气息，又融入了时尚的全球文化。右图将视角锁定在一条里弄的入口，透过那被排排灯笼映衬着的饰品及美食商铺，带你走进田子坊的日常，原汁原味地还原其中西合璧的特色。

Tianzifang

The tiger and bear warmly welcome you to Tianzifang! Here, you will experience the essence of Shanghai culture as you witness the convergence of local lifestyles and global cultural trends. As you walk down the boulevards of Tianzifang, you can also enjoy the numerous antique shops, coffee houses, art studios, and restaurants.

黄浦区泰康路 210 弄
Lane 210 Taikang Rd., Huangpu District

- 田子坊所处的泰康路艺术街 1998 年前为马路集市，在区政府实施马路集市入室后，以画家陈逸飞为代表的艺术家、设计师的进驻使泰康路开始被改造成艺术街

- 田子坊曾为上海石库门里弄、厂房等，经改造成为艺术气息浓厚的工作室、工艺商品店、画廊、咖啡馆等；小巷错综相连，少有死胡同

- 名字来源于画家黄永玉几年前给这旧弄堂起的雅号，取中国古代画家"田子方"的谐音

- Taikang Road, the street next to Tianzifang, was originally a street market, and was transformed into an art street after designers and artists such as Chen Yifei moved in since 1998.

- Art studios, shops, galleries, and coffee shops adapted from small factories, warehouses, and houses; alleys all narrowly connected and branching off each other; one can rarely find dead ends

- Named by the Chinese artist Huang Yongyu after the ancient Chinese artist Tian Zifang

武康大楼

　　高耸挺立的桅杆，蓄势待发的大炮，乘风破浪的船体——武康大楼不再只是一栋普通的公寓，而是成为了一艘无往不胜的战舰！在现实中，这座建筑从西侧看确实像极了一艘威风凛凛的战舰。赶快提起画笔，在澎湃的涛声中勇往直前，继续探险之旅吧！

Normandie Apartments

　　Are you wondering why the Normandie Apartments is sailing on the sea? Do the mast, cannon, and hull on the building remind you of a battleship taking off for combat? These fictional features highlight how the building, now also called "Wukang Building" in Chinese, was once said to be nicknamed after the battleship *Normandie*. Pick up your pen and start coloring to embark on your own voyage in the roaring sea!

- 始建于 1924 年

- 万国储蓄会出资，邬达克设计

- 平面以楔状地形布置成熨斗形，30 度锐角头部位于五条马路交会的岔路口。立面纵向三段式划分——底层为商铺，拱廊式骑楼；顶层由连通的挑出阳台和女儿墙构成双重水平线脚的檐部。三层有山花装饰的窗楣

- 沪上最早的外廊式公寓；曾名"I. S. S. 公寓"（I. S. S. 为万国储蓄会的英文简称）、诺曼底公寓等；大楼电梯仍保留着老式半圆指针型楼层指示器

- Built in 1924

- Commissioned by International Savings Society; designed by László Hudec

- Iron-shaped layout; apartments with exterior corridors; stone railings on the terrace; pediment decorations on lintels of windows

- Originally named I. S. S. Apartments, it was the oldest exterior-corridor apartment in Shanghai. Now it is called Wukang Building in Chinese due to its location on Wukang Road.

徐汇区淮海中路 1842-1858 号（近武康路）
No. 1842-1858 Middle Huaihai Rd. (near Wukang Rd.), Xuhui District

小红楼

咿咿呀呀，呀呀咿咿，随着那老式留声机再度唱起时代的沧桑沉浮，你以为你穿越回了老上海的考究人家，殊不知来到的其实是坐落于衡山路、半掩在绿地翠枝波影间的"小红楼"。小红楼曾被法国百代唱片公司打造成"东方百代唱片公司"，右图将它作为上海第一座录音棚的前世映刻在古旧的留声机中，而它作为"小红楼"西餐厅的今生则可从图中精致的红酒杯与蕾丝桌布中窥见一斑。

Little Red House

If you see a gramophone, you may feel like you've travelled back to the 1920s. In fact, you can see one in the Little Red House located in Xujiahui Park. The intricately designed music box in the shape of the House on a restaurant table reflects the building's history. Originally the best and largest recording studio in China, it is now the Western restaurant La Villa Rouge. Its name corresponds with its signature red-brick exterior.

徐汇区衡山路 811 号（徐家汇公园内）
No. 811 Hengshan Rd. (in Xujiahui Park), Xuhui District

- 建于 1921 年
- 1908 年法国"百代公司"在上海成立中国第一家唱片公司——东方百代唱片公司，1921 年购下今小红楼所在地，建起上海第一座录音棚
- 孟莎式坡屋顶，老虎窗，檐下承以牛腿木托架，隅石贴面
- 聂耳在此创作了《义勇军进行曲》，至今仍广为传唱的《玫瑰玫瑰我爱你》《夜来香》等 20 世纪 30 年代流行歌曲也诞生于此

- Built in 1921
- Established as a recording studio by French company Pathé Records
- Red brick walls and white cement decorations in the exterior; mansard roof with roof windows
- In this building, the *March of the Volunteers* was produced, which later became the national anthem of the People's Republic of China.

新天地

"昨天，明天，相会在今天"——当你信步走在新天地那青砖步行道上，穿梭于青红相间的清水砖墙间，你仿若置身于20世纪二三十年代的旧上海；而当你一步跨入建筑内，一间间装修精致、极具情调的高档餐厅、咖啡馆则又将你拉回当下，继续着现代都市人的生活方式与节奏。右图以典型的石库门门楣作为框架，透过门洞可见诸多现代休闲场所，希望将中与西、过去与当下完美结合。

Xintiandi

You can witness the combination of Western and Chinese cultures in Xintiandi. The Shikumen townhouses were originally built by Western settlers in Shanghai during the Chinese civil war from the 1850s to 1860s. Looking through the Shikumen, you can see the modern restaurants and buildings that occupy Xintiandi today. The intricate carvings on the gate are also key highlights of Shikumen.

黄浦区黄陂南路、自忠路、马当路、太仓路围合区域
Area enclosed by South Huangpi Rd., Zizhong Rd., Madang Rd., and Taicang Rd., Huangpu District

- 原为20世纪初形成的本地居民住宅，20世纪末由香港瑞安集团改造，1999年动工，2001年建成

- 分为南里和北里——南里以现代建筑为主，北里则以保留石库门旧建筑为主；石库门特色体现为厚重的乌漆大门和雕着巴洛克风格卷涡状山花的门楣

- 保留原有砖、瓦作为建材；改造为由明星经营的餐饮娱乐中心、画廊、邮局博物馆等休闲场所

- Formed residential area at the beginning of 20th century; redeveloped and completed in 2001

- Shikumen (stone-framed gate) townhouses; lilong (alleys); lintels sculpted with Chinese characters or rococo motifs

- The Xintiandi area was renovated by Hong Kong's Shui On Group. The original bricks and tiles were well preserved during the renovation. Now it is a major cultural and entertainment complex with shops, cafes, and restaurants.

徐家汇天主堂

有"远东第一大教堂"之称的徐家汇天主堂是上海现存唯一的双钟塔式哥特教堂,两侧高达56米的对称尖塔可谓是建筑的一大特色。你可有看出图中仿教堂穹顶的八边形就是由多个标志性的尖塔所组成的呢?而在这众多尖塔指向的中心肃立着的就是基督教的心脏——耶稣。

St. Ignatius' Cathedral

This illustration highlights the bell towers, which are a distinctive feature of the St. Ignatius' Cathedral. Depicted here are the bell towers rearranged to form the typical shape of a church dome. Just like how Jesus Christ is the central figure of Christianity, you can find the statue of Christ in the center of the main façade of the cathedral — and this illustration.

徐汇区蒲西路 158 号
No. 158 Puxi Rd., Xuhui District

- 1910 年竣工

- 由建筑师道达尔设计,旨在服务于 18 世纪日渐增多的江南天主教友

- 平面为拉丁十字型,主立面向东,由两侧高达 56 米的尖塔和中间后退的山墙组成;外部结构采用清一色红砖,墙基部分则为青石;门窗都是哥特尖拱式,嵌彩色玻璃;堂内的每根楹柱均由 10 根小圆柱拼合,是西式柱式的中国做法;侧廊彩绘玻璃上画有圣经中的场景及传统中式元素

- "远东第一大教堂",大厅可容纳 2 500 余人;江南教区的中心,有"远东梵蒂冈"之称;著名导演斯蒂芬·斯皮尔伯格 1987 年的作品《太阳帝国》曾在此取景拍摄

- Completed in 1910

- Designed by British architect W. M. Dowdall

- Masonry structure with twin spires; red brick walls; white stone pillars; bell towers; colorful stained glass windows; cross-shaped plan

- The St. Ignatius' Cathedral displays a combination of Chinese and Western decorations on windows at one side of the corridor — classic stories from the Bible are painted in the middle, typical Chinese patterns such as abacus and rockeries on top, and Chinese illustrations of Bible stories below. Steven Spielberg's movie *Empire of the Sun*, set in WWII Shanghai, was filmed here.

杨树浦水厂

　　晨间洗漱、晚间淋浴时，你是否有过疑问：这哗哗的水流究竟是从何而来的呢？如果你住在上海，那么坐落于黄浦江边的杨树浦水厂也许就是答案——毕竟这座全上海最老的自来水厂可是为整座城市提供了四分之一的水源呢！享受涂色的同时，可也别忘了把正在滴水的水龙头拧紧哦！

Yangshupu Water Plant

　　When you brush your teeth in the morning or take a shower at night in Shanghai, do you ever wonder where all that tap water comes from? It may very well be from the Yangshupu Water Plant, Shanghai's oldest water plant. The valves and the faucet with dripping water make the entire building seem like a huge water tank, signifying upon the building's character as a water plant. Remember to turn off the faucet after you wash your hands!

- 最初由英国商人创办，1883年建成供水，后陆续扩建

- 扩建部分承重墙用清水砖墙，嵌以红砖饰带，门窗洞采用哥特式尖拱，屋顶四周筑锯齿状矮墙，形似英国城堡

- "远东第一大水厂"；中国第一座现代化水厂；李鸿章拧开阀门开闸放水标志着水厂的建成

- Established by Britsh enterprises in 1883, extended afterwards

- Classical British castle style; pointed arches, crenellated roofs; blue brick walls with red brick decorations

- The Yangshupu Water Plant is the first modern water plant in China and is dubbed the "first water plant of the Far East." It once provided up to 1.53 million cubic meters of water a day.

杨浦区杨树浦路830号（近许昌路）
No. 830 Yangshupu Rd. (near Xuchang Rd.), Yangpu District

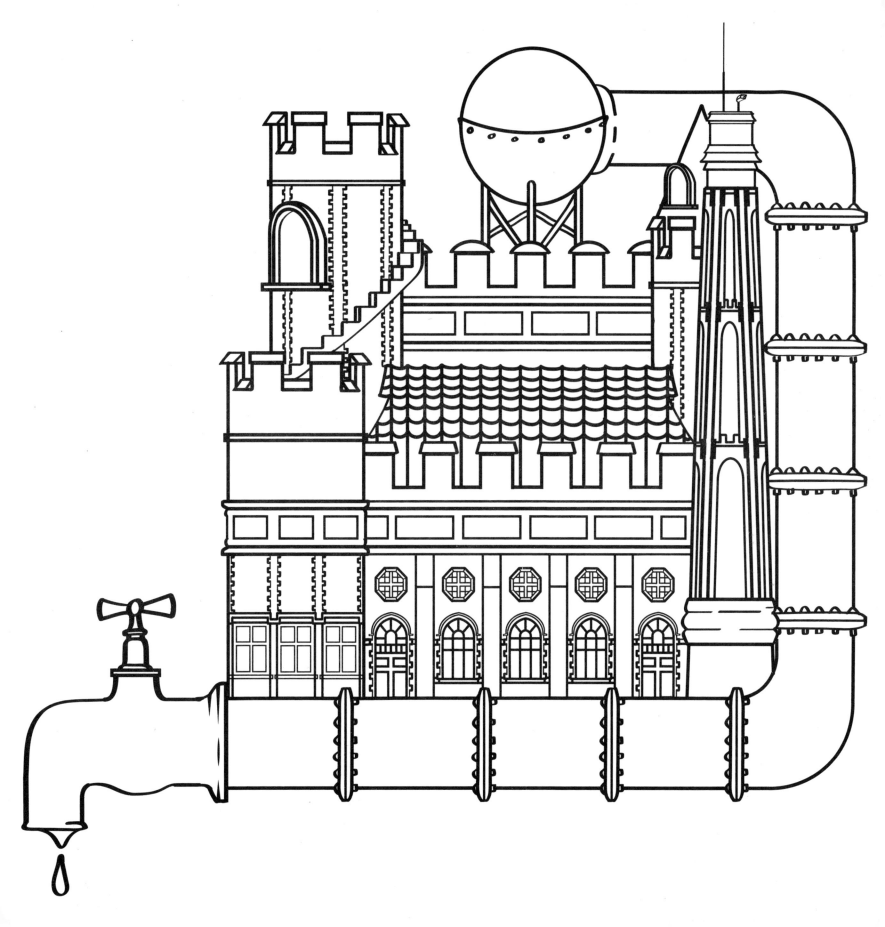

豫园

　　都市的生活偶尔太过喧嚣忙碌，如果想要在一个安静的午后，去往古色古香的亭台楼阁让自己浮躁的心沉静下来，位于上海老城区的豫园一定是一个不错的选择。为了体现出江南古典园林的建筑特色，图中豫园有着密布的花草树木、嶙峋的假山奇石和错综相连的楼阁；而图中央那婉约静美、身着明代襦裙的女子则标志着豫园始建于明代的历史。

Yuyuan Garden

　　If you are stressed out by the bustling city life in Shanghai, a visit to the old pavilions at Yuyuan will help you find some relaxation. The alluring woman dressed in Ming dynasty clothing alludes to Yuyuan's origin, as it was first built in the Ming dynasty. The pavillion on the woman's head, exquisite rockeries, and the abundant flora surrounding her are highlights of Yuyuan.

黄浦区安仁街
Anren Rd., Huangpu District

- 始建于明代

- 明朝人潘允端聘请造园名家张南阳设计建造

- 因其出色的景色、布局与规模，建成当时即被公认为"东南名园冠"；潘允端在《豫园记》中注明"豫"有"平安安泰"之意，取名"豫园"则为"豫悦老亲"的意思

- 屡遭破坏，几经改建，清末曾为兵营；新中国成立后数次修缮，1982年被定为全国重点文物保护单位；收藏众多名石、泥塑、书画、木刻等历史文物；园内举办各类花展、灯会、茶道等活动；周边市场成为"中国四大文化市场"之一

- Built in Ming dynasty

- Created as a private garden of Pan Yunduan, designed by gardening expert Zhang Nanyang

- The name "Yu" means "peace" or "soundness" in Chinese. It is widely recognized as the "best garden in the Southeast" with numerous intricate rockeries, lush trees, ponds, bridges, pavilions, and terraces.

- As a National Cultural Heritage Conservation Unit, Yuyuan contains many historical artifacts like calligraphy, paintings, and clay sculptures. It also hosts activities such as tea ceremonies, lantern shows, and flower exhibitions.

朱家角

说起江南水乡，人们脑海中定会浮现出一幅小桥流水人家的静美画面。位于上海边界的朱家角古镇不仅风景如画，还蕴涵着1 700年的历史。如今这里不仅吸引了来自五湖四海的游客，还成为许多外来人口包括少数民族同胞创业发展的热土。逆流而上，依稀仿佛间，你可见那窈窕淑女在水中伫立？轻言细语间，你可有从她那别具特色的头饰和服装中，看出她来自于哪一少数民族？

Zhujiajiao

When you think of a Jiangnan water town, you may envision small bridges, flowing brooks, and village households. You can travel to the outskirts of Shanghai to Zhujiajiao, where you can find exactly these. A historic Chinese water town with more than 1 700 years of history, Zhujiajiao is home to many ethnic minorities today. In this illustration, you can see a Zhuang minority woman whose dress is morphed into water, adding to the picturesque scenery.

- 1 700年前三国时期，朱家角已形成村落集市；元代，镇上建起圆津禅院、慈门寺等古寺名刹，人丁集居，初具规模；明代正式建镇

- 九条老街依水傍河，千余栋民宅临河而建，36座古桥古风犹存，粉墙灰瓦曲径通幽

- 号称"衣被天下"，以布业著称江南；明末清初，朱家角米业突起，带动百业兴旺，故又有"三泾（朱泾、枫泾、泗泾）不如一角（朱家角）"之说；上海市四大文化名镇之一

- Formed water town around 1 700 years ago

- Criss-crossed river ports; village streets stretched along the river; Ming and Qing dynasty architecture; ancient stone bridges

- Zhujiajiao is a typical Jiangnan style water town with charming natural scenery, often called the "Venice of China." It also has numerous river ports, teahouses, and well-preserved pavilions and other buildings from the Ming and Qing dynasties. In 1991, Zhujiajiao was designated as one of the four cultural towns of Shanghai by the Shanghai Municipal Government.

青浦区朱家角古镇（G50 朱家角出口）
Zhujiajiao Ancient Town (G50 Expressway Exit Zujiajiao), Qingpu District

趣味问答

涂色的间隙，来做几道选择题吧，看看你是哪幢建筑的真爱粉？（答案在书内找哦！）

1. 武康大楼曾名？
   A. I. S. S. 公寓
   B. 邬达克公寓
   C. 熨斗大楼
   D. 徐汇公寓

2. 以下关于鸿德堂正确的是：
   A. 前身是录音棚
   B. 有着模仿中式庙堂的建筑风格
   C. 形似城堡
   D. 为上海提供大量水源

3. 以下关于丁香花园正确的是：
   A. 它建于明代
   B. 花园结构形似丁香
   C. 有着蜿蜒的龙墙
   D. 比豫园年代久远

4. 1933老场坊原本是一座：
   A. 大学
   B. 屠宰场
   C. 图书馆
   D. 监狱

5. 以下关于豫园正确的是：
   A. 位于上海老城区
   B. 建造年代在丁香花园之后
   C. 融入了斯堪的纳维亚式的建筑风格
   D. 受西方建筑风格影响

6. 位于新天地的民居被称为：
   A. 弄堂
   B. 摩天大厦
   C. 邬达克
   D. 石库门

7. 以下哪座建筑的屋顶是一座花园？
   A. 绿房子
   B. 江湾体育场
   C. 国际饭店
   D. 小红楼

8. 以下哪座建筑原为中苏友好大厦？
   A. 小红楼
   B. 百乐门舞厅
   C. 华东政法大学
   D. 上海展览中心

9. 以下哪处位置建有石库门？
   A. 豫园
   B. 丁香花园
   C. 外滩
   D. 田子坊

10. 以下关于马勒别墅正确的是：
    A. 形似战舰
    B. 原为中苏友好大厦
    C. 是一座外观似城堡的私人住宅
    D. 有着迷宫般的结构

11. 华东政法大学的前身是：
    A. 圣约翰大学
    B. 圣约翰图书馆
    C. 圣约翰教堂
    D. 以上都不正确

12. 截至20世纪80年代，上海最高的大楼是以下哪一座？
    A. 诺曼底公寓
    B. 小红楼
    C. 鸿德堂
    D. 国际饭店

13. 百乐门舞厅有着以下哪种特征？
    A. 蜿蜒的龙墙
    B. 雕刻着动物的外墙
    C. 圆柱形的霓虹灯
    D. 屋顶花园

14. 以下关于江湾体育场正确的是：
    A. 由一位英国富翁建造
    B. 曾是一座展览馆
    C. 有配套体育馆和游泳池
    D. 由邬达克设计

15. 大光明电影院由以下哪个特征著称？
    A. 圆顶结构
    B. 横竖交错的线条
    C. 对称的尖顶
    D. 江南园林元素

16. 以下关于杨树浦水厂正确的是：
    A. 曾是屠宰场
    B. 现在是休闲娱乐中心
    C. 是中国第一座现代化水厂
    D. 为上海提供过半的水源

17. 小红楼的前身是：
    A. 录音棚
    B. 水厂
    C. 拘留所
    D. 监狱

18. 以下关于朱家角正确的是：
    A. 是典型的中国水乡
    B. 形成于18世纪晚期
    C. 由石库门建筑群组成
    D. 受斯堪的纳维亚建筑风格的影响

19. 徐家汇天主堂具有以下哪个特征？
    A. 对称的尖顶
    B. 中式庙堂元素
    C. 迷宫般的内里结构
    D. 圆顶

Time for an ArchiQuiz!

Congratulations! You finished coloring all of the buildings. Now, find out how much you have learned about Shanghai's architecture! (Answers can all be found in this book.)

1. The Normandie Aprtments was once named?
   A. I. S. S. Apartments
   B. Hudec Apartments
   C. Flatiron Building
   D. Xuhui Apartments

2. The Fitch Memorial Church ____
   A. was originally a recording studio
   B. is built in Chinese temple style
   C. looks like a castle
   D. is a major water supplier in Shanghai

3. The Lilac Garden ____
   A. was built during the Ming dynasty
   B. looks like a lilac
   C. features zigzag dragon walls
   D. was built earlier than Yuyuan

4. The 1933 Old Millfun was originally used as ____
   A. a university
   B. a slaughterhouse
   C. a library
   D. a prison

5. The Yuyuan Garden ____
   A. is located in the Old City of Shanghai
   B. was built after Lilac Garden
   C. incorporates Scandinavian style
   D. was heavily influenced by the West

6. The houses in Xintiandi are called ____
   A. longtang
   B. skyscraper
   C. Hudec
   D. shikumen

7. Which of these buildings has a garden roof?
   A. the Green House
   B. Jiangwan Sports Stadium
   C. Park Hotel
   D. Little Red House

8. Which of these buildings was originally the Sino-Soviet Friendship Building?
   A. Little Red House
   B. Paramount Hall
   C. East China University of Political Science and Law
   D. Shanghai Exhibition Center

9. Which of these features shikumen?
   A. Yuyuan Garden
   B. Lilac Garden
   C. The Bund
   D. Tianzifang

10. The Moller Villa ____
    A. looks like a battleship
    B. was originally built as a Sino-Soviet Friendship building
    C. is a castle-like private residence
    D. has a maze-like structure

11. The East China University of Politics and Law was originally ____
    A. St. John's University
    B. St. John's Library
    C. St. John's Church
    D. None of the above

12. Which was the highest building in Shanghai until the 1980s?
    A. Normandie Apartments
    B. Little Red House
    C. Fitch Memorial Church
    D. Park Hotel

13. The Paramount Hall features ____
    A. zigzag dragon walls
    B. animal murals on the exterior
    C. cylindrical-column shaped neon tower
    D. rooftop flower garden

14. The Jiangwan Sports Stadium ____
    A. was commissioned by a British magnate
    B. used to be an exhibition center
    C. was an outdoor field designed along with an indoor stadium and a swimming pool
    D. was designed by L. Hudec

15. The Grand Theatre is characterized by ____
    A. dome structure
    B. horizontal and vertical lines
    C. symmetrical spires
    D. Jiangnan style garden houses

16. The Yangshupu Water Plant ____
    A. used to be a slaughterhouse
    B. is now an entertainment complex
    C. is the first modern water plant in China
    D. supplies water to more than half of Shanghai

17. Little Red House used to be ____
    A. a recording studio
    B. a water plant
    C. an internment camp
    D. a prison

18. Zhujiajiao ____
    A. is a traditional Chinese water town
    B. formed in the late 1700s
    C. features townhouses called shikumen
    D. influenced by Scandinavian architectural styles

19. The St Ignatius' Cathedral features ____
    A. symmetrical spires
    B. Chinese style temple structures
    C. maze-like interior
    D. dome structure

找一找，这些小图标分别和哪些建筑有关？

Each icon is associated with one building or place in this book. Can you figure out which?

上海还有哪些有趣的建筑？
提起你的画笔，发挥你的想象力吧！

Now search for a piece of architecture in Shanghai and add your own artistic interpretation!

推荐阅读
Recommended Readings

1. 上海建筑施工志编委会编写办公室.东方"巴黎":近代上海建筑史话 [M].上海:上海文化出版社,1991.

2. 上海市旅游局.经典上海建筑之旅:英文 [M].上海:上海文化出版社,2011.

3. 姜庆共,席闻雷.上海里弄文化地图:石库门 [M].上海:同济大学出版社,2012.

4. 华霞虹,乔争月,齐斐然,卢恺绮.上海邬达克建筑地图 [M].上海:同济大学出版社,2013.

5. 周进.上海教堂建筑地图 [M].上海:同济大学出版社,2014.

6. 王唯铭.墙·呼啸:1843年以来的上海建筑 [M].上海:文汇出版社,2007.

7. 诺凡利.上海——有着中西百年建筑的城市 [M].上海:百家出版社,2002.

8. 蔡育天.回眸:上海优秀近代保护建筑 [M].上海:上海人民出版社,2001.

9. 陈从周,章明,上海市民用建筑设计院.上海近代建筑史稿 [M].上海:上海三联书店,2002.

10. 惜珍.花园洋房的下午茶:上海的保护建筑 [M].上海:东方出版中心,2010.

11. 何巍,朱晓明.上海工部局宰牲场建筑档案研究 [J].时代建筑,2012(3):108-113.

12. 赵崇新.1933老场坊改造 [J].建筑学报,2008,(12):70-75.

13. 妍丽,张健."中西合璧"在上海近代独立式花园住宅中的体现 [J].华中建筑,2007,25(4):116-118.

14. 姚素梅.上海近代独立式住宅花园研究(1840-1949)[D].上海交通大学,2009.

15. 劳旺,劳叶.辉煌在历史的细节里:纪念上海展览中心(原中苏友好大厦)建成50周年 [EB/OL]. ABBS建筑论坛,2005-03-21 [2017-01-03]. http://www.abbs.com.cn/jzsb/read.php?cate=5&recid=12731.

16. 徐汇区湖南路街道办事处.武康大楼口述史项目专题片 [Z/OL].酷6网,2016 [2017-01-03] http://v.ku6.com/show/ab8itf61LRmglEpWKGhRvg...html.

17. Mu Qian (China Daily). Shanghai's charm revealed [EB/OL].中国日报网,2011-10-27 [2017-01-03]. http://www.chinadaily.com.cn/cndy/2011-10/27/content_13984230.htm.

18. 应忠德.杨树浦水厂档案接收日志 [EB/OL].上海档案信息网,2013-04-24 [2017-01-03]. http://www.archives.sh.cn/zxsd/201304/t20130424_38396.html.

19. 乔晓红.历史地段建筑环境的再生与创新:记上海太平桥地区新天地广场旧城改建项目 [J].建筑学报,2001(3):12-15.

20. 上海市房产经济学会卢湾分会.卢湾区太平桥地区旧区改造启示 [J].上海房地,2001(11):34-35,48.

21. Chinadiscover.net. Shanghai Xintiandi [EB/OL].中国日报网,2012-10-08 [2017-01-03]. http://www.chinadaily.com.cn/life/2012-10/08/content_15801128.htm.

22. 上海地方志办公室
   www.shtong.gov.cn

23. Hudec Heritage Project
   www.hudecproject.com;
   epiteszet.hudecproject.com

24. 上海1933老场坊官网
   www.1933shanghai.com

25. 华东政法大学
   www.ecupl.edu.cn

26. 上海新天地
   www.shxintiandi.com

27. 上海豫园
   www.yugarden.com.cn

28. 朱家角古镇旅游门户站
   www.zhujiajiao.com

一本涂色书的诞生

在制作《多彩城市》的一年多里，我们常常问自己：我们的初衷是什么？

这一切都要由2015年12月的中国大智汇挑战赛说起。我们虽然从小在上海长大，却因长久在国际学校接受西式教育而逐渐疏远了上海的传统文化。这让我们意识到现在年轻一代普遍缺乏对上海文化的了解，所以我们决心改善这一问题。

我们以沪上青少年对上海建筑文化的认知作为切入点，呼吁人们重拾那些街道楼宇所承载的海派文化——因为我们认为上海的特色建筑是其作为海派城市最好的代表。而"中国大智汇"，作为一个鼓励中国高中生通过团队合作去解决当代社会的诸多问题的创新研究比赛，为我们实现这一愿望提供了一个很好的平台。

作为参赛者，我们采用了查阅文献、分发问卷、实地考察、采访专业人士等多种方法进行研究，每一阶段的研究成果都以线上互动的形式发布在"阿拉桑海宁"的微信公众平台上。三个月后，我们最终绘制了一本长达52页的原创双语建筑涂色本，以留白的手绘形式呈现了20处具有历史意义和文化价值的上海建筑，希望以此能够让人们在一个富含创意与互动的过程中，学习上海建筑并提高文化意识。之后，我们联系了同济大学出版社，得到了出版发行我们涂色本的机会。

最终，我们有幸获得了2016"中国大智汇"全国第二名，这一份沉甸甸的荣誉也让我们决心进一步扩大课题的影响力。在同济大学出版社各位专业老师的指导与建议下，我们对涂色本进行了多方面的修缮，注入了更多原创性的元素。在一次次的协商后，最终完成了各位读者手中正捧着的这本书——《多彩城市》。

回首过去的一年多，我们曾跌跌撞撞地经历许多困难。作为高二、高三的学生党，最大的挑战必然是保持学业和课外活动之间的平衡。在无数个被作业、论文、截稿日期折磨的夜晚，我们都咬着牙坚持了下来。另一大难题是在制作涂色本的过程中，我们需要在趣味性和教育意义之间进行取舍。在编辑文字时，我们既要确保建筑信息的准确性，又要避免文字太过枯燥，影响到涂鸦的感受和读者对建筑的兴趣。在几番讨论后，我们最终决定在以趣味小贴士的形式科普建筑相关知识的同时，加入基于我们自身对建筑理解的创意元素。此外在选择建筑的图片作为手绘与图标的原型时，我们需要确保这些图片不仅具有代表性，也要适合读者上色。我们还需要维持每幅建筑手绘"写实性"与"创意性"的平衡，在保留作画人原创灵感的同时，不能让建筑与其真实外形构造偏离太多。

其实，通过制作这本涂色本，我们自己也收获了许多。在一年多的合作中，我们成为了彼此更好的队友和搭档——尽管有困难和争执，但本着共同的奋斗目标，我们学会了去聆听、去妥协、去信任。

《多彩城市》不仅凝聚了我们的心血、蕴含了我们对上海的爱，更见证了一路走来我们的成长。我们虽然不一定有最精美的插图和最专业的介绍来让所有人都对上海建筑无所不知、无所不晓，但是我们希望能够培养一种兴趣、一种意识、一份对脚下这片土地所承载的文化历史的欣赏与尊重。我们由衷地希望在读完、涂完《多彩城市》后，您对上海的爱，也能如我们的一般，从心底漫溢至这城市的每一个街角。

The Making of *City of Colors*

It all started when we read about China Thinks Big (CTB), a social entrepreneurship competition for high school students to conduct their own research. Aware of the fading presence of Shanghai's architecture in local culture, we realized that this platform was a great opportunity for us to begin our project and contribute to our community.

With their motto "Think Big, Do Small" in mind, we completed our research through reading reference materials, interviewing experts and peers, distributing surveys, and conducting on-site investigations. Eventually, we utilized all the information we obtained to develop a bilingual coloring book containing hand-drawn illustrations of 20 buildings with brief explanations of their cultural and historical significance. We believe that such a coloring book allows both children and adults to participate in an engaging and creative learning process as opposed to the traditional methods of memorization and recitation.

After such efforts, we were honored to be awarded Second Place on a national scale by the CTB committee. At the same time, we contacted Tongji University Press and received the wonderful chance to publish it officially. Under the guidance of editors at Tongji University Press, we made several revisions and modified our book into what you're holding right now.

As you have seen in the previous pages, the illustrations of the architecture have been infused with elements of our imagination, and you may be wondering why. Our primary purpose for creating this book was to raise awareness of Shanghai architecture, so we wanted to make sure that our book was of educational value. At the same time, we wanted to make our book interesting and easy to color for our readers. To meet both our visions, we eventually decided to illustrate realistic architectural features with a creative twist and also incorporate factual information from our research into our text.

Needless to say, the year-long journey of producing *City of Colors* was no easy feat for us. After all, even though we were curious and passionate towards Shanghai, we were still clueless high school students when we first started our project. One of our greatest challenges was managing our school work and research at the same time, considering we were in our busiest two years of high school. We spent countless nights researching Shanghai architecture and adjusting our illustrations while our teachers pushed us to turn in our homework, essays, and lab reports. Reflecting on our process, we believe what enabled us to accomplish this project was undoubtedly our dedication and persistence.

Other notable challenges include selecting 19 iconic buildings and transforming them into creative illustrations. Initially, we considered many buildings before narrowing our list down to the ones most representative of the diverse architecture in Shanghai. Then, we looked over countless photos to ensure an accurate but also creative portrayal in our illustrations and icons. For some buildings, creative inspiration for our designs came naturally, such as Moller Villa, which immediately reminded us of a castle in fairytales; but for others, suitable concepts came only when we attempted multiple ideas. When drafting Jiangwan Sports Stadium, for instance, we tried many concepts such as tessellating sports equipment and depicting the stadium from different angles before deciding on our final design.

Although the process was challenging, we are proud to say that our time and effort were not wasted. Producing *City of Colors* allowed us to not only grow as individuals — developing our problem-solving skills, research abilities, team spirit, and creativity, but it also gave us the opportunity to express our appreciation for Shanghai. We may not be professional writers or artists, but we sincerely wish that this book has increased your interest for the history and culture of Shanghai. We hope that you now love Shanghai as much as we do and will perhaps even choose to do more exploration of this beautiful city on your own!

致 谢

回首这一年多的时光，我们想感谢的人有许多：

感谢上海中学国际部——作为我们成长道路上的第二个家，培育出了今天的我们。

感谢张丽老师——带领并指导我们参加比赛，不断给予我们鼓励与支持；

感谢阮仪三教授、华霞虹副教授以及斯文·赛拉诺老师——接受我们的采访并给予我们宝贵的建议和参考资料；

感谢中国大智汇——为我们提供这样一个平台，使我们能够有机会去完成这一份对我们来说意义非凡的课题；

感谢杜超瑜女士——对书中的建筑知识进行校核及补充；

感谢同济大学出版社的江岱女士、罗璇女士、朱笑黎女士以及其他参与出版工作的各位老师——在我们对出版这一行业完全不熟悉的情况下，带领我们进行一系列的出版工作并耐心解答我们的各种疑问；

感谢所有支持我们的家人、老师和朋友——不论是帮助我们填写问卷、转发我们公众微信号的帖子，还是在上海书展现场为我们宣传捧场，你们所做的一切我们都铭记在心，感激不尽！

Many Thanks To...

We would like to express our immense gratitude to the many who have made the publication of this book a reality:

SHSID, our beloved school, for being our second home and cultivating us into who we are today.

Ms. Zhang Li, our supervisor, for guiding us through the project and giving us constant encouragement;

Professor Ruan Yisan, Associate Professor Hua Xiahong, and Mr. Sven Serrano, for helping us with our research and providing us with useful resources;

The Harvard College Association for US-China Relations (HAUSCR) for organizing China Thinks Big 2016 and giving us the opportunity and platform to embark on this meaningful project;

Ms. Du Chaoyu for proofreading the architectural facts in this book;

Ms. Jiang Dai, Ms. Luo Xuan, Ms. Zhu Xiaoli, and many others of Tongji University Press for bearing with us through our endless questions as we tread into the publishing business for the first time;

And last but not least, our friends, teachers, and families for putting up with our numerous surveys and promotions and giving us the greatest support we could have ever asked for!

图书在版编目（CIP）数据

多彩城市：涂出你的上海 = City of Colors:
Bring Shanghai's Architecture to Life! : 汉英对照 /
阿奇上海著. -- 上海：同济大学出版社，2017.5
　ISBN 978-7-5608-6854-7

Ⅰ．①多… Ⅱ．①阿… Ⅲ．①建筑物－介绍－上海－
汉、英 Ⅳ．① TU-862

中国版本图书馆 CIP 数据核字（2017）第 070214 号

多彩城市：涂出你的上海
City of Colors: Bring Shanghai's Architecture to Life!

阿奇上海　著
ArchiShanghai

出 品 人：华春荣
策划编辑：江　岱
责任编辑：罗　璇
责任校对：徐春莲
装帧设计：孙晓悦
出版发行：同济大学出版社　www.tongjipress.com.cn
地　　址：上海市四平路 1239 号　邮编：200092
电　　话：021-65985622
经　　销：全国各地新华书店
印　　刷：上海安兴汇东纸业有限公司
开　　本：787mm×1092mm　1/12
印　　张：5
字　　数：126 000
版　　次：2017 年 5 月第 1 版　2019 年 3 月第 2 次印刷
书　　号：ISBN 978-7-5608-6854-7
定　　价：30.00 元